The First Book of Electronics Workshop: Can't Beat A Practical Approach!

RIVER PUBLISHERS SERIES IN COMMUNICATIONS

Consulting Series Editors

MARINA RUGGIERI
University of Roma "Tor Vergata"
Italy

HOMAYOUN NIKOOKAR
Delft University of Technology
The Netherlands

This series focuses on communications science and technology. This includes the theory and use of systems involving all terminals, computers, and information processors; wired and wireless networks; and network layouts, procontentsols, architectures, and implementations.

Furthermore, developments toward newmarket demands in systems, products, and technologies such as personal communications services, multimedia systems, enterprise networks, and optical communications systems.

- Wireless Communications
- Networks
- Security
- Antennas & Propagation
- Microwaves
- Software Defined Radio

For a list of other books in this series, visit www.riverpublishers.com
http://riverpublishers.com/river publisher/series.php?msg=Communications

The First Book of Electronics Workshop: Can't Beat A Practical Approach!

Dr. Bhawani Shankar Chowdhry

Dr. Ahsan Ahmed Ursani

Muhammad Zaigham Abbas Shah

MUET, Pakistan

River Publishers

Aalborg

Published, sold and distributed by:
River Publishers
Niels Jernes Vej 10
9220 Aalborg Ø
Denmark

ISBN: 978-87-93102-47-7 (Print)
      978-87-93102-48-4 (Ebook)
©2014 River Publishers

Contents

Circuit Simulation

Appendices

Index   **83**

Authors Biography   **85**

Preface

The field of electronics has seen an unparalleled growth in the last 60 years, from the invention of the transistor to the making of the processor. In this ever evolving field, the modern day student has been observed to jump to complex circuit designing without having a firm understanding of the internal circuit elements and the tools that are used to analyze them. This book is an attempt to redress these shortcomings by providing an apt and concise description of basic electronic components and apparatus and how to work with them practically. Theoretical description is followed by specifying the practical considerations so as to cement the student's understanding of the component/apparatus. This edition contains a more detailed component description with focus on real life usability. We have included many pictures showing the different shapes and forms of each component that are available. A set of questions has been included after each practical so as to challenge the students understanding of the component discussed. Tasks have been changed so they relate more to everyday situations and build up student intuition. A section on working with components has been included which introduces the student to basic circuit elements that can be made using various components. Discussion is also done on noting and analyzing various phenomenon's that occur during circuit operation such as phase difference etc.

The Practical Book on Electronic Workshop imparts technical knowledge on following five main topics:

Laboratory Apparatus

Passive Electronic Components

Active Electronic Components

Circuit Assembly

Circuit Simulation

It is envisaged that before students use any of the lab equipment for conducting any practical work, they must become familiar with their use and functions.

Similar is the case with the passive and active electronic components. The students mostly perform their practical work in the senior semester over specialized trainers, and never get acquainted with practicality of circuit components. Hence they face severe problems while working over their own projects. Similarly, knowing how to build circuits is as important as knowing how to design circuits and how to use the components. Therefore, techniques of Circuit Assembling are also covered in this Practical Book.

Though this book adopts practical approach, it first gives a thorough and sound theoretical background of each and every apparatus and component covered in the book, and then it reinforces the theoretical concepts by discussing their practical considerations. We feel that this Practical Book on Electronic Workshop is first of its kind, and will find usefulness for the students of all engineering disciplines in general, and Electrical, Electronics, Telecommunication in particular.

We believe that this Practical Book will be valuable and insightful in getting basic knowledge and skills of Exciting and Important field of Electronics.

AA Ursani, Zaigham Abbas Shah, BS Cowdhry

Safety Precautions

1. Never hurry. Work deliberately and carefully.
2. Do not touch components which are hot with your bare hands.
3. Ensure components are powered in the correct required polarity.
4. Before using any component make sure to read its rated voltage and do not exceed its power ratings.
5. All conducting surfaces intended to be at ground potential should be connected together.
6. Do not run wires or rotating equipment, or on the floor, or string them walkways from bench-to-bench.
7. Check circuit power supply voltages for proper value and for type (AC, DC, frequency) before energizing the circuit.
8. Connect to the power source last.
9. When breaking an inductive circuit, open the switch with your left hand and turn your face away to avoid danger from any arc which may occur across the switch terminals.
10. In case there is smoke or a burning smell coming from a circuit, disconnect the power source first.
11. In case you find any naked live wire, power cable of any equipment without its proper plug, immediately inform the Lab assistant.
12. Remove conductive watch bands or chains, finger rings, wrist watches, etc, and do not use the metallic pencils, metal or metal edge rulers, etc. when working with exposed circuits.
13. Keep liquids away from electrical circuits/equipment.
14. In case you find any fire, immediately inform the Lab assistant.
15. Do not touch the soldering iron to check whether it is hot enough.
16. When soldering components on to Veroboards/PCB, ensure that the soldering iron is placed in its stand to avoid getting burned.

Workshop # 01

Power Supplies and Measuring Instruments

Object: To become familiar with various types of Power Supplies and Electrical Measuring Instruments.

Apparatus: Power Supply units and Measuring Instruments in the Workshop

Theory
Power Supplies

A power supply unit is a source of either a Constant Current or Constant Voltage irrespective of the load resistance. Power sources can be classified as either varying or non varying, i-e AC or DC. AC stands for Alternating Current where as DC stands for Direct current. Graphically speaking, the Alternating currents are those that vary in some periodic fashion electrically and reverse polarity several times. The voltage which causes an Alternating Current is called AC voltage. Figure 1.1 shows the wave shape of an alternating power source.

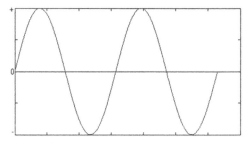

Figure 1.1   The AC power source

The sinusoidal waveform is the most popular, mostly because it is the waveform generated by alternators. Consequently, it is the form that comes in our house mains. It has a magnitude of 220V and a frequency of 60Hz, meaning that the voltage alternates polarity 60 times per second. The value 220V is the RMS (Root Mean Square) value of the incoming mains supply.

The RMS value of an AC voltage is the equivalent DC energy delivered to the load.

The voltage that causes Direct Current is called DC Voltage. Unlike its AC counterpart, Direct Current flows with the same polarity at all times. An example of a Voltage source would be that of a battery or our home mains supply. Figure 1.2 shows the response of an ideal DC power source.

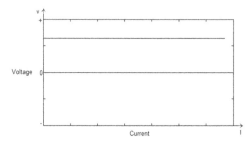

Figure 1.2   The ideal DC power source

The curve shown in Figure 1.2 is for an ideal source. A real power source behaves a little differently. The curve for a real source shown in Figure 1.3:

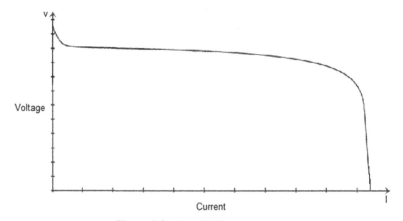

Figure 1.3   A real DC power source

As seen in the Figure 1.3, the ideal voltage source is supposed to provide the same voltage for an infinite amount of time, but for a real voltage source (e.g. a battery), the voltage begins to droop as more and more current is taken from the source. This is because a voltage source (the battery), cannot contain an infinite amount of charge. The same is true for a current source too. Figure 1.4 shows the symbols used for various types of voltage and current sources.

Symbols of some Current/Voltage Sources are:

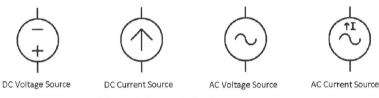

DC Voltage Source   DC Current Source   AC Voltage Source   AC Current Source

Figure 1.4   Symbols of power sources

Measuring Instruments

Measuring instruments allow us to get an estimate of some physical quantity. An instrument that can measure Potential Difference (voltage) is known as a Voltmeter, one that measures current is known as an Ampere Meter (Ammeter for short) and resistance is measured using an Ohmmeter. An instrument which can measure all three quantities is known as a Multimeter or sometimes an AVO (Ampere-Volt-Ohm) Meter. Traditionally all of the instruments consisted of a dial and a deflection pointer and thus were called Analog instruments. These devices required the correction of errors which showed up in them from time to time. This process of correcting the error is known as Calibration. Now digital instruments have replaced their old analog counterparts, unnecessitating the process for calibration. Also, Digital Instruments display readings on a Liquid Crystal Display making them easier to read from as compared to analog instruments which often consist of multiple and/or non linear scales.

Voltmeters have high impedance and hence are connected in parallel across the points between which potential difference is to be measured. The reason that they have high impedance is that they do not steal away any current from the measured circuit. Ammeters, on the other hand exhibit a very small resistance, and hence are connected in series, they should never be connected in parallel since they may short out the circuit and result in it getting damaged. Figure 1.5 illustrates how to connect Voltmeters and Ammeters in a circuit.

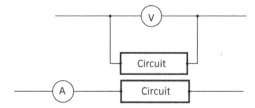

Figure 1.5   Connecting a voltmeter and an ammeter in a circuit

Multimeters usually have a selectable switch which allows you to select the quantity that you want to measure and the maximum value called Range, of the quantity that you want to measure as shown in Figure 1.6:

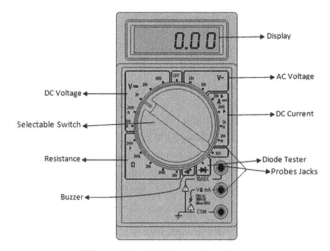

Figure 1.6   Multimeter functions

Procedure

1. Look for of Power Supply Units of any kind in the Workshop, like AC Current Source, AC Voltage Source, DC Current Source, DC Voltage Source, and list them all in Table 1.1. Note down their Type, Model, and the Range. Some of them have a digital display; some of them might have a scale with a pointing needle.
2. Look for various kinds of Measuring Instruments e.g. Voltmeter, Amme-ter, Ohmmeter etc, available in the Workshop. Note down their Model

No's, Type and the Quantities which they can measure and the range of measurement of each quantity in Table 1.2.

3. Take any voltage source, connect a Voltmeter to its output terminals and turn it ON. Select any voltage from the voltage source, note down the selected value from its scale or display, against the value that you measure, in Table 1.3. Is the value exactly the same as selected? Repeat the same by selecting different magnitudes of voltage each time.

4. Finally, try measuring the mains AC Voltage. Take a Digital Multimeter (DMM), and set it to measure AC voltage. Select the measurement range higher than 220 Volts, and insert the DMM probes in anyone of the HALF POINTS in the Workshop.

Observations

1. List few of the Power Supply Units available in the Workshop.

Table 1.1   Determining the ranges and types of power supplies

S. N.o	Model	Type (Tick the Relevant Choice)				Range
		AC	DC	Voltage	Current	
1.						
2.						
3.						
4.						
5.						

2. List few of the Measuring Instruments available in the Workshop.

Table 1.2   Determining the ranges and types of measuring instruments

S. N.o	Model	Type (Tick the Relevant Choice)		Measurement						Range
				Quantities						Range
		Analog	Digital	V AC	I AC	V DC	I DC	R	Other	
1.										
2.										
3.										
4.										
5.										

3. Fill in the following table for any of the Voltage Sources

Table 1.3   Determining the error in voltage output of a voltage source

S N.o	Selected Voltage	Observed Voltage	Percent Error
1.			
2.			
3.			
4.			
5.			

Questions

1. What is the difference between a Voltage Source and a Current Source?

   _____

   _____

2. Write down the voltage and current ratings of a few appliances in your home.

   _____

   _____

3. How can one use the multimeter to find shorts in a circuit.

   _____

   _____

4. Fill in the table for the value of the quantity provided the Range is chosen as given in the Table 1.4

   Table 1.4   Checking effect of multimeter range on display of measurements

S No	Range	Reading on Multimeter	Value of quantity
1	10KΩ	2.5	
2	100Ω	25	
3	200 mV	20.00	
4	2V	0.02	
5	20μA	7.8	
6	10mA	0.2	

5. Describe a simple method of increasing the range of Ammeters and Voltmeters.

   _____

   _____

Workshop # 02

Oscilloscope and Function Generators

Object: To become familiar with Oscilloscopes and Function Generators

Apparatus: Oscilloscope and function generator

Theory

Electronic circuits need to be tested for their functionality. This is done by applying test signals as input to the circuits and observing their response. This requires a device which generates test signals and another device for observing the circuit response to those test signals.

A Function Generator is a device that is used to produce (generate) test signals for testing electronic circuits. These can generate a number of periodically varying electrical signals. The Front Panel of the function generator is shown in Figure 2.3. These are one of the most versatile sources available, allowing us to manipulate several parameters such as:

1. Function or wave shape- All function generators can generate three basic types of waveforms i-e Sinusoidal, Triangular and Rectangular. They can also produce composite waveforms which are combinations of any two or all the three waveforms. This can be done using the waveform shape select control. Figure 2.1 shows the basic waveshapes that can be generated by a function generator.

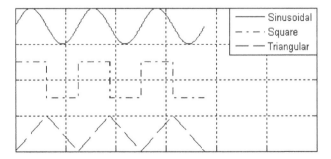

Figure 2.1   The sinusoidal, square and triangular waveforms

2. Time Period or Frequency- A wide range of frequencies can be generated (0.01 Hz to 10 MHz or even more). Adjusting the Frequency Multiplier and the Frequency Selector knob helps select the frequency of the generated signal.
3. Amplitude- The amplitude of the output signal can be increased and decreased over a wide range using the attenuation control.
4. Dc Level- If one wants to add a DC offset to our signal so as to move the signal above or below the ground line than this can be done using the DC Offset control. This is shown in Figure 2.2.

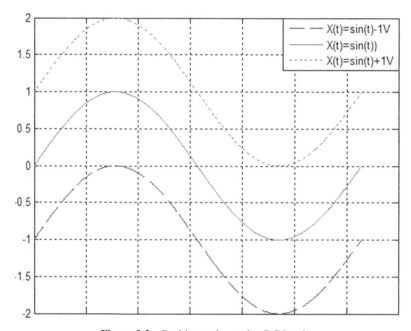

Figure 2.2   Positive and negative DC levels

5. Frequency Sweep- Function Generators also allow us to sweep the frequency both linearly and logarithmically, this option is available in the new digital function generators.
6. Modulated Signals- Digital Function Generators allow us to produce Amplitude, Frequency and Pulse Amplitude Modulated signals as well.

Function generators also have probe connections for taking the signal outputs.

An Oscilloscope (or simply a Scope for short) is a device that lets you observe circuit responses in form of waves or oscillations on a screen and

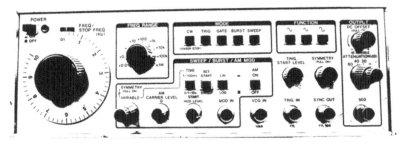

Figure 2.3 Function generator front panel

possibly make measurements as well. The graph that appears against a grid on the screen is called a Trace. The Horizontal and Vertical axes represent Time and Amplitude respectively. Each square on the grid is called a Division and this is used to calculate the amplitude as well as time period of the signal. Each division is further divided into 5 subdivisions with each subdivision representing 0.2 divisions. Most oscilloscopes consist of two inputs called Channels, which help you observe two signals simultaneously. Each channel has the following controls:

1. Volts/Division- This is used to scale the signal so that it can be accommodated within the screen and no part of the signal goes beyond the display area. This is also known as the Vertical Scale. The amplitude of the signal is then given by:

   Amplitude(p–p)= Number of vertical blocks covered by signal x Vertical Scale

2. Coupling- There are three ways in which the signal can be coupled to the channel, namely GND, AC and DC as shown in Figure 2.4.

   i. GND- This lets you see the ground (reference) line for the signal.

   ii. AC- This lets you AC couple the signal i-e any DC component in the signal is filtered out before displaying it on the screen.

   iii. DC- This lets you DC couple the signal i-e displays the complete signal including its AC as well as DC part.

3. Vertical Position- This lets you move the waveform up and down on the screen. It is used for setting the ground line on the origin.

4. Input Mode- This control is common for both the channels, this lets you choose the input channel or combination mechanism for the display. It consists of four options, CH1, CH2, Alternate, Chop and Add. When this control is on Alternate, the channels are displayed alternately in

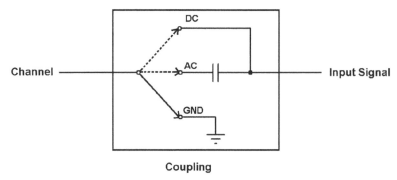

Figure 2.4   Coupling in an oscilloscope

successive sweeps, in the Chop mode, both the channels can be viewed simultaneously and the Add mode displays the sum/difference of the two channels.

5. Invert- This is present on only one of the channels (mostly Channel 2), this inverts the displayed signal. It is used for subtracting one signal from the other using the Add input mode.

6. X-Y Mode- Most oscilloscopes also have an X-Y mode option that lets you plot one channel against the other. This can be used to plot I-V curves.

7. 10x- The probes for each channel have a 10x switch on their heads, which divides the signal by 10 before displaying it on the scope thus allowing for much larger signals to be displayed on the screen.

The controls used for manipulating the horizontal axis of the signal are:

1. Time/Division- This is similar to the Volts/Division control described before, and lets you scale the time period of the signal i-e either compress it or stretch it horizontally. The time period of the signal is then given as:

Time Period = Number of horizontal blocks covered by one cycle x Horizontal Scale

2. Horizontal Position- This lets you move the trace either to the left or to the right. This is used to position the waveform in such a way so that its starting point touches the origin.

3. X10 Magnifier- This allows you to measure a frequency 10 times higher than the normal range of the oscilloscope. This is used to measure time deflections.

The controls discussed above were used to change the waveform on the screen. The oscilloscope uses a horizontal sweep to catch the input signal after equal intervals of time thus making the signal seem stationary on the screen (ideally). This is done with the help of the trigger controls:

1. Level and Slope- This lets you select the amplitude and the slope (+ve or -ve) at which to begin the sweep.

2. Mode- This consists of the Normal, Auto and Single Sweep modes, the normal mode produces a sweep only when the signal (from the source selected) crosses the set value of Level moving in the direction of the Slope. The Auto mode produces a free running sweep. The Single mode is used to produce only one cycle of a sweep and is generally used for non repetitive signals.

3. Source- This switch lets you choose the source for the trigger, the options are CH1, CH2, Line, Ext and Ext÷10. The CH1 and CH2 select either Channel 1 or Channel 2 for the trigger input, the Line input causes the sweep to be triggered on the AC power line, Ext and Ext÷10 causes it to trigger on any external signal.

4. Coupling- The coupling switch lets you select the coupling used for the trigger source; this consists of DC and AC options. The AC and DC options, respectively, AC couple and DC couple the trigger source.

An oscilloscope also has Intensity (make the trace brighter/dull), Focus (change the focus of the trace), Probe connectors (BNC connectors for connecting the probes), a Cursor Function (lets you measure the amplitude, time period, frequency, phase difference etc automatically) and a Calibration Output. The intensity should not be very high nor the focus be made blurry as it could result in the damage of eyesight. The calibration Output lets you Calibrate the oscilloscope. Before making any observation with the Oscilloscope, one must calibrate both the channels to make sure that the observations are correct to the maximum possible accuracy. The oscilloscope generates a signal called the Calibration Signal, which is a square wave of 1 KHz i-e a Time Period of 1ms. This signal is observed on both the channels one by one, any error in the observation is corrected before making any other observation. There is a limit to the highest frequency that can be displayed on an oscilloscope. Most of the oscilloscopes can display signals with a maximum frequency of 20 MHz. Figure 2.5 shows the front panel of the Oscilloscope.

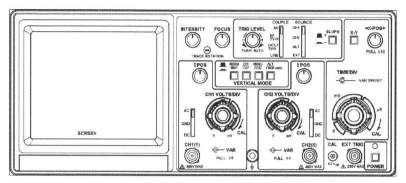

Figure 2.5   Oscilloscope front panel

Procedure

1. Turn on both pieces of equipment.
2. Plug the probes to the probe inputs on the oscilloscope.
3. Select Channel 1 and select GND from the Channel Coupling switch, if it is not already on the origin, use the vertical position knob to bring the Ground line down to the origin.
4. Repeat the above step for Channel 2 also.
5. Connect the probe of Channel 1 to the calibration output. Use the volts/division switch to set the vertical scale to 0.5 volts/division.
6. Remove the probe of Channel 1 form the calibration output and repeat the same for Channel 2.
7. To calibrate the time scale, connecting any of the probes of Channel 1 or Channel 2 to the calibration output, use the horizontal position switch to move the wave horizontally so that its starting point rests on the origin.
8. Set the Vertical Coupling switch to AC, connect the probe of Channel 1 as well as Channel 2 to the function generator output.
9. Press the Sine key on the function generator and set the frequency multiplier knob to 1 and the frequency selector knob to 1 KHz.
10. Use the attenuation switch on the Function Generator to set the amplitude to 2 Volts (p–p), while observing on the oscilloscope.
11. Measure the amplitude and the time period using the formula given in theory. Use the cursor function and measure the amplitude and the time period and fill in Table 2.1.
12. Do this for several frequencies.
13. Press Alt or Chop on the vertical coupling switch and view both the waveforms simultaneously.

14. Change the Channel mode to add and observe the waveform on the screen.
15. Press the Invert control and observe the change.
16. Observe different shapes of waveforms that the function generator can produce by using the Waveform shape switch. Also observe the waveform by pressing two or more of the buttons together.
17. Set the Coupling mode to DC. Using the DC offset knob on the function generator, lift the signal level up by 2 divisions on the oscilloscope. Now switch the coupling mode to AC and observe the difference.
18. Repeat the above step but move the signal 2 divisions down instead.
19. Take another function generator, press the sine key and while observing on the oscilloscope, set the frequency to 1 KHz and the amplitude to 2 Volts (p–p).
20. Now observe the signal from the first function generator on channel 1 and the signal from the second generator on channel 2 (chop mode).
21. Now calculate the time difference between (using the block method) the starting points of both the waveforms and use the formula described below to calculate the phase difference between the two waves as shown in Figure 2.6. You will also need to calculate the period of the waves.

$$Phase\ Difference = \frac{td}{T} \times 360^o$$

Observation

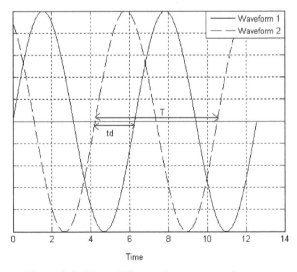

Figure 2.6   Phase difference between two sine waves

Table 2.1   Determining Inaccuracy in the generated frequency

S N.o	Function Generator Frequency	Observed Frequency		Error between F.G Freq. and Calcu-lated Freq.	Error between F.G Freq. and Cursor Func. Freq.
		Calculated	Cursor Function		
1.					
2.					
3.					
4.					
5.					

Questions

1. Draw the waveform generated when the sinusoidal and rectangular waveforms are combined.

   _____

   _____

2. What is the purpose of the volts/division and the time/division control?

   _____

   _____

3. What does one mean by coupling?

   _____

   _____

Workshop # 03

Resistors

Object: To become familiar with resistors, variable resistors and resistor color coding

Apparatus:

1). A Digital Multimeter (DMM)
2). Few fixed Resistors
3). Few Variable Resistors

Theory

A Resistor is a component used to reduce the amount of current passing through a circuit. The measure of the capability of the resistor to limit the current is known as Resistance and its unit is *Ohm*. Mathematically, Resistance is given by Ohm's Law:

$$V = IR \text{ or } R = \frac{V}{I}$$

Where V is the voltage applied across the material and I is the current flowing through it. Resistors are mostly made from poor conductors, the most common material being Carbon film or Metal film, low value resistors are however made by wires wound over ceramic former. There are some special types of resistors which are made from semiconductors. Resistors come in two types, fixed or variable. Fixed resistors, as the name suggests are the ones whose resistance cannot be varied where as the resistance of variable resistors can be changed. The symbols for fixed resistors are shown in Figure 3.1.

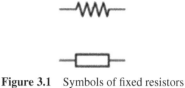

Figure 3.1   Symbols of fixed resistors

The resistance is not the only thing to be taken in to consideration when choosing a resistor, the tolerance and the power rating also need to be considered. Tolerance is the maximum deviation a resistance (or any other component for that matter) can exhibit. The power rating indicates how much power the resistor can safely withstand.

Fixed resistors come in standard values which are specified by the manufacturer using colored bands made on the resistor itself, there are three band standards, the 4–Band, 5–Band and the 6–Band standard as shown in Figure 3.2.

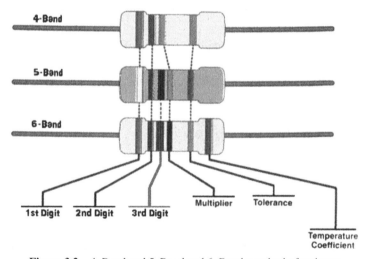

Figure 3.2   4–Band and 5–Band and 6–Band standard of resistors

As seen in Table 3.1, Gold and Silver are not used in either the Units, Tens or Hundreds bands, also, not all the colors are used in the tolerance band. As an example, if a 4 band resistor has the following color codes, Brown in 1st, Black in 2nd , Red in the 3rd and Gold in the 4th, than its resistance is going to be:

Table 3.1 Color codes to resistor bands

Color	Value	Multiplier	Tolerance
Black	0	0	-
Brown	1	1	±1%
Red	2	2	±2%
Orange	3	3	±0.05%
Yellow	4	4	-
Green	5	5	±0.5%
Blue	6	6	±0.25%
Violet	7	7	±0.1%
Grey	8	8	-
White	9	9	-
Gold	-	-1	±5%
Silver	-	-2	±10%

Brown	Black	Red	Gold
1	0	$x10^2$	±5%

Its minimum and maximum resistance can be calculated using the tolerance value:

Minimum Resistance: $\left(1000 - \frac{1000 \times 5}{100}\right)\Omega$

$$(1000 - 50)\Omega$$

$$950\Omega$$

Maximum Resistance: $\left(1000 - \frac{1000 \times 5}{100}\right)\Omega$

$$(1000 + 50)\Omega$$

$$1050\Omega$$

Resistors come in various series (classified according to the tolerance levels) such as E6 (20% Tolerance), E12 (10% Tolerance) and E24 (5% Tolerance). Precision Resistors having tolerances of 1% are also available but they are expensive as compared to the commonly available 5% and 10% Tolerance resistors.

Variable resistors consist of a slider by which one can change their resistance. There are two types of variable resistors, the Potentiometer and the Rheostat. The potentiometer has 3 terminals where as the Rheostat has

2. These find applications as dimmers in fans, volume control in radios etc. Figure 3.3 shows the symbols of variable resistors:

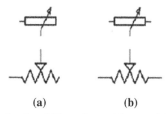

(a)   (b)

Figure 3.3   Symbols of variable resistors (rheostat (a), potentiometer (b))

Some potentiometers have values written on them, as shown Figure 3.4 on others though, the value needs to be decoded.

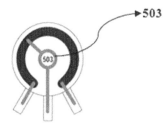

Figure 3.4   Potentiometer outlook

As seen in Figure 3.4, the number 503 is written on it, its value can be calculated by using the formula:

$$(1^{st} \text{ Digit})(2^{nd} \text{ Digit}) \times 10^{Multiplier}$$

So, its value becomes

$$50 \times 10^3 \Omega \text{ or } 50000 \Omega$$

Some Fixed and Variable Resistors have values written on them, the suffixes that they use are shown in Table 3.2.

So if a resistor has the value 20E0, it means its value is 20Ω and 2K2 would mean 2.2$\times$1000Ω or 2200Ω.

Power ratings of resistors used in common electronics circuits are 1/8, 1/4, 1/2 and 1 Watts, higher ratings are also available. One can use the following formula for finding out an appropriate power rating for a specific resistor value:

$$P = I^2 R$$

Table 3.2 Suffix meanings

Suffix	Denotation
R,E	Decimal Point
K	Times 1000
M	Times 1000000

Where P is the power dissipated by the Resistor, I is the current flowing through it and R is the value of the Resistance. As a rule of thumb, choose a resistor having power rating equal to twice the calculated value. Some real resistors are shown in Figure 3.5.

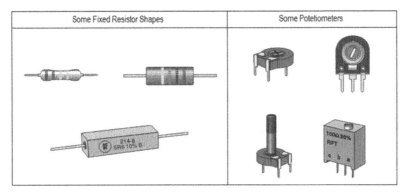

Some Fixed Resistor Shapes	Some Potetiometers

Figure 3.5   Some real resistors

Procedure

1. Take any Resistor and decode its Resistance from the Color Bands (CR).
2. Take a Multi-meter. Select the option of Ohmmeter with appropriate range.
3. Place the probes on the terminals of the Resistor to measure the Resistance (MR).
4. Calculate the Percent Error as *(CR − MR)*x100/*CR* and write in Table 3.3.
5. If the Percent error is less than the Tolerance, the Resistor is reliable and is in accordance with specifications.
6. Repeat (1) through (6) four times.

7. Take a Potentiometer and connect its two end terminals with the Ohmmeter.
8. Turn the knob of the Potentiometer in any direction, while observing its Resistance.
9. Connect the probe of the Ohmmeter across the middle terminal of the Potentiometer and any one of the end-terminals.
10. Turn the knob of the Potentiometer fully clockwise and note down its Resistance.
11. Turn the knob of the Potentiometer slowly in counter-clockwise direction while observing its Resistance on the Ohmmeter.

Observations

Table 3.3   Difference between rated and real value of a Resistor

S. N.o	Resistance		Percent Error	Tolerance	Implication
	Coded	Measured			
1.					
2.					
3.					
4.					
5.					

Questions

1. Write down the values of the following resistors in ohms:
   i. 2M3   ii. 5E4   iii. 94R2   iv. 7K2

   _____

   _____

2. Can the Potentiometer be used as a Rheostat?

   _____

   _____

3. Write down a few applications of Variable Resistors?

   _____

   _____

Workshop # 04

Capacitors

Object: To become familiar with Capacitors, their types and reading their values

Apparatus:

1). A Digital Multimeter (DMM)
2). Few Electrolytic and Non Electrolytic Capacitors

Theory

A capacitor is an electronic component used to store electric charge. Together with resistors and inductors, it is the most frequently used component in electronics. Capacitors are made up of two metallic plates having an insulator (also called a Dielectric) in between them, which enables it to store charges in the form of an electric field. The kind of dielectric used along with several other factors, determines how much charge the capacitor can store. Whenever the terminals of a capacitor are connected across a battery, there is a deficiency of electrons on one plate and an excess of electrons on the other. This creates a potential difference between the two plates and gives rise to an electric field. The capacity of a Capacitor to store charges is known as its capacitance which has the unit Farad (F), however, Farad is typically a big unit and one usually talks about capacitance in much smaller units such as µF, nF and pF etc.

There are two types of capacitors, Electrolytic and Non Electrolytic (symbols shown in Figure 4.1).

(a)   (b)

Figure 4.1 Symbol of electrolytic (a) non-electrolytic (b) capacitors

Electrolytic capacitors are polarized in that they need to be connected the correct way round (+ve supply to +ve terminal and –ve supply to –ve terminal)

to charge them. They come in large as well as small values but are usually greater than 1µf. They come in two main types; one consists of metal foils with oxide insulators (electrolytic) and hence is simply called an Electrolytic Capacitor. The other type of polarized capacitors is Tantalum capacitors. The positive end is marked by a "+" and the negative lead is marked by a "–". Since these capacitors are usually physically large, their values are printed on them along with their voltage ratings and tolerances. Electrolytic capacitors are used for filtering out ripples in DC power supplies.

Non-Electrolytic capacitors on the other hand can be charged with any polarity and are thus non-polarized. They use Mica, Glass, Paper, Ceramic, Porcelain, Polycarbonate and Wax as the dielectric and are usually less than 1µf. They are usually used in AC circuits along with resistors and inductors to perform mathematical operations and filtering.

Ideal capacitors have infinite resistance. Real capacitors show a very high resistance in the order of 100's of Kilo Ohms, in fact this is one way to check whether a capacitor is faulty or not. When connected across a multimeter, with the range set to measure up to 1MΩ, a working capacitor would show zero first and gradually rise to a very high value (this is because the capacitor is being charged by the battery of the multimeter), if it is faulty, than it will stay at zero and the capacitor is said to have become shorted, if the capacitor has become opened, there will be no reading on the multimeter.

Electrolytic capacitors, as mentioned before have values printed on them, on the other hand, the values of non electrolytic capacitors have to be decoded. This takes in to consideration a general rule plus some replace by consideration with regard to the value non electrolytic capacitors can have i-e <1µF. The general rule is

$$(1^{st}\text{Digit})(2^{nd}\text{Digit}) \text{x} 10^{\text{Multiplier}} \mu\text{F/pF}$$

Determining the unit is where the consideration kicks in, say if a capacitor has the number 102 written on it, and this is how one would decode its value,

$$10 \text{x} 10^2 \text{pF} = 1000\text{pF} = 1\text{nF}$$

We took the unit to be pF because if we considered it to be µF, the capacitance would come out to be 1mF which would be too large a value for non-electrolytic capacitors. As another example, let's say that the number written on the capacitor is .02, now the value of this capacitor would be decoded as .02µF since .02pF is too small a capacitance value.

Some capacitors also have several suffixes following the coded values, the symbols for the tolerances and their corresponding values are shown in Table 4.1.

Table 4.1   Capacitor tolerance codes and values

Tolerance Code	Tolerance Value
Z (For Large capacitors)	+80%, –20%
M	±20%
K	±10%
J	±5%
G	±2%
F	±1%
D	±0.5%
C	±0.25%
B	±0.1%
A	±0.05%
Z (For Small Precision Capacitors)	±0.025%
N	±0.02%

The capacitors discussed so far were fixed value capacitors; there are also variable capacitors that allow us to vary the capacitance. This is achieved in two ways, in one method, there is a set of fixed (stator) plates and some movable (rotor) plates in between those stator plates, the movable plates can be brought into and taken out of the mesh by means of a shaft thus varying the capacitance, in the next method one has a mechanism for moving the dielectric thus changing the capacitance. The capacitance of variable capacitors ranges between 1pF and 500pF. Variable capacitors of very small value are known as Trimmer capacitors. These are used in radios for tuning purpose. Symbols are shown in Figure 4.2.

Every capacitor has a voltage rating for which it can be used, if the capacitor is supplied with a voltage greater than the rating, it will result in it getting damaged. Some real capacitors are shown in Figure 4.3.

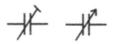

Figure 4.2   Symbol of variable (a) trimmer (b) capacitors

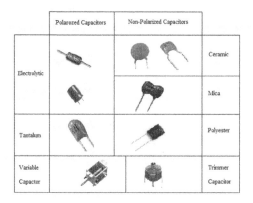

Figure 4.3   Some real capacitors

Procedure

1. Take a Multimeter that can measure capacitance, set it to measure resistance with the range set at 2MΩ.
2. Take a few capacitors (polarized and non polarized). Taking each capacitor one at a time, use the method described in the theory to check whether the capacitor is faulty or not and write down your observation in Table 4.2.
3. Now set the multimeter to measure capacitance, decode/read the capacitance value and write it down in Column 3 of Table 4.3 measure the capacitance with the multimeter and write down in the table as well.

Observations

Table 4.2   Checking a capacitor

S N.o	Polarized/Non-Polarized	Value	Implication Faulty or not faulty
1.			
2.			
3.			
4.			
5.			

Table 4.3 Tolerance checking of a capacitor

S N.o	Polarized/Non-Polarized	Value	Decoded/Read Capacitance	Measured Value	Difference	Percent Error
1.						
2.						
3.						
4.						
5.						

Questions

1. What happens to the overall capacitance if we connect two capacitors in series or in parallel?

   _____

   _____

2. Decode the following capacitance values:
   i. 104   ii. 560   iii. 474

   _____

   _____

3. Write down some of the applications in which capacitors are used.

   _____

   _____

Workshop # 05

Inductors and Transformers

Object: To become familiar with inductors and transformers

Apparatus:

1). A Digital Multimeter (DMM)
2). Few inductors
3). Few transformers

Theory

An inductor is a coil of wire usually wound over a Ferro-magnetic material which can store energy in the form of a magnetic field. The ability of an inductor to store energy is given by its Inductance having the unit Henry (H). Once again Henry is a very large unit and we usually talk about inductance in mH or μH. The symbol of an inductor is shown in Figure 5.1.

ΛΥΥΥ

Figure 5.1   Symbol of an inductor

An ideal inductor would exhibit zero resistance but a real inductor has some finite resistance. Inductors which have become open will show infinite resistance, whereas shorted inductors will show a resistance of zero Ohms. Although one can wind inductors, conventional inductors come in a resistor like shape with color bands on them. The procedure to decode inductor values is shown in Figure 5.2.

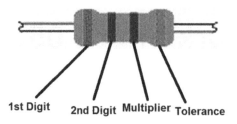

Figure 5.2   Inductor color bands

$$(1^{st}\text{Digit})(2^{nd}\text{Digit}) \times 10^{(\text{Multiplier})} \mu H$$

The color codes are given in Table 5.1.

Table 5.1   Color codes for Inductors

Color	Value	Multiplier	Tolerance
Black	0	0	±20%
Brown	1	1	±1%
Red	2	2	±2%
Orange	3	3	±3%
Yellow	4	4	±4%
Green	5	-	-
Blue	6	-	-
Violet	7	-	-
Grey	8	-	-
White	9	-	-
Gold	-	−1	±5%
Silver	-	−2	±10%

For example if an inductor has the band color sequence Blue, Grey, Black and Silver, the value of the inductor would be decoded as,

Blue	Grey	Brown	Gold
6	8	x 10^1	±5% µH

So, the rated value of the inductor is 680 µH, its minimum value can be 646 µH and its maximum value can be 714 µH. An inductor is like an electromagnet, when supplied with a current, it energizes up thus creating a magnetic field, in fact this is the reason they find applications in Speakers, Bells etc.

Variable inductors are those inductors whose inductance can be varied. This is accomplished either by a movable core or by having a movable contact that can be moved along the coil. Both these mechanisms are housed inside a small casing. These are used in radios along with capacitors for tuning purpose. The symbol of the variable inductor is shown in Figure 5.3.

Figure 5.3   Symbol of a variable inductor

A transformer is a device that uses a pair of coils/windings to manipulate AC voltage and current. It uses the principle of mutual induction to either decrease or increase the voltage/current at its input. The input winding is known as Primary winding and the Output winding is known as Secondary winding. Transformers are used at Powerhouses to increase the voltage for transmission and near homes to decrease it to a usable value. The voltage/current at the output of the transformer is given by the following equation:

$$\frac{Vs}{Vp} = \frac{Ns}{Np} = \frac{Ip}{Is}$$

Where V_s, V_p, I_s and I_p represent Primary Voltage, Secondary Voltage, Secondary Current and Primary Current respectively. The symbol of a transformer are shown in Figure 5.4.

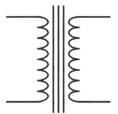

Figure 5.4   Symbol of a transformer

These are two types of transformers

1. Step Up Transformer: Increases the voltage and decreases the current
2. Step Down Transformer: Decreases the voltage and increases the current

Step Up transformers have more number of turns in the Secondary winding as compared to the Primary winding while the opposite is true for Step Down transformers. This can also be checked by measuring the resistance. For a step up transformer, the primary (greater no of turns) winding resistance would be greater than the secondary winding resistance and vice versa. There is also a special type of transformer in which there a wire that protrudes out from the middle of the secondary coil thus allowing for the secondary voltage to be taken in two halves. This type of transformer is known as a Centre-Tapped transformer (shown in Figure 5.4.).

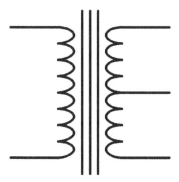

Figure 5.5   Symbol of a Centre-Tapped transformer

There is another type of transformer known as the Linear Variable Differential Transformer. The secondary winding of this transformer consists of two sections each wound in the opposite direction as shown in Figure 5.6.

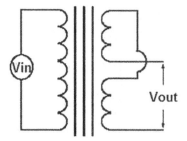

Figure 5.6   Symbol of an LVDT

This is used for measuring displacement. Some real inductors are shown in Figure 5.7.

Fixed Inductors	Variable Inductors

Figure 5.7 Some real inductors

Procedure

1. Pick up an inductor.
2. Setting the multimeter to measure Resistance with a range of 200, find out whether the inductor is faulty or not.
3. Pick up a few transformers. Set the multimeter to measure voltage with a range of 400V AC. Measure the voltage in the mains socket and write down in the Table 5.2. Now connect the transformers primary winding to the mains socket.
4. Set the multimeter to a range of 50V, measure the voltage at the secondary of the transformer and write down in the table. Use the formula given in the theory to find out the number of turns of the transformer.
5. Repeat steps (3) and (4) for a few more transformers.

Observation

Table 5.2 Determining the turn ratio of a transformer

S N.o	Peak to Peak Input Voltage	Output Voltage	Turn Ratio
1.			
2.			
3.			
4.			
5.			

Questions

1. What happens to the overall inductance if we connect two inductors in series or in parallel?

   _____

   _____

2. Decode the following inductor values:
   i. Red Violet Brown Gold

   _____

   _____

   ii. Yellow Blue Brown Gold

   _____

   _____

3. What does one mean by inductive kick?

   _____

   _____

4. Comment on the role of the Inductor in an LC tank circuit.

   _____

   _____

5. Enlist some applications of a transformer?

   _____

   _____

6. Describe briefly (in terms of the turn ratio and purpose) the functioning of the Current and Potential transformers?

   _____

   _____

Workshop # 06

Switches and Relays

Object: To become familiar with various type of switches

Apparatus:

1). A Digital Multimeter (DMM)
2). Few hard wires (Gauge Number 22)
3). A Few switches
4). 2 Lamps
5). A 5V, 12V DC Power Supply

Theory

A switch is an electromechanical component that can be used to electrically connect or disconnect two points in a circuit. Switches are of many kinds, some have a mechanical actuating mechanism (hand switches), some work using a magnetic field (relays) while some even get turned on using light or temperature etc. Switches can be classified according to a number of criteria. In terms of the number of contacts, switches can be classified in to 3 types:

1. Single Pole Single Throw Switch (SPST)
2. Single Pole Double Throw Switch (SPDT)
3. Double Pole Double Throw Switch (DPDT)

Poles and Throws are the two sets of contacts on any switch, by virtue of its actuating mechanism, a switch connects poles and throws to each other and is said to be *Closed*, when the poles and throws are not connected to each other, the switch is said to be *Open*.

1. Single Pole Single Throw Switch (SPST):

A Single Pole Single Throw switch is the simplest type of switch available in that it consists of only two terminals which are connected together whenever the switch is actuated. Such types of switches are used for home appliances. Figure 6.1 shows the symbol of an SPST switch. Figure 6.2 shows the connection of a SPST with a lamp.

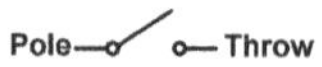

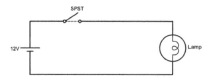

Figure 6.1   Symbol of an SPST Switch

Figure 6.2   Connecting an SPST switch to a lamp

2. Single Pole Double Throw Switch (SPDT):

A Single Pole Double Throw switch consists of three terminals of which one is a pole and the other two are throws. At any time the pole can only be connected to one of the throws. Such switches are used in changeovers for home mains supply. Figure below shows the symbol of an SPDT switch. Figure 6.4 shows the connection of a SPDT to two lamps.

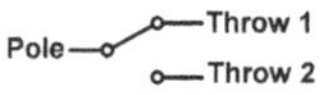

Figure 6.3   Symbol of a SPDT switch

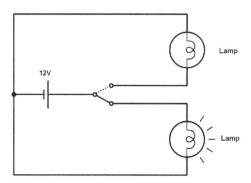

Figure 6.4   Connecting an SPDT switch to two lamps

3. Double Pole Double Throw Switch (DPDT):

A Double Pole Double Throw switch consists of six terminals of which two are poles and the rest are throws. It is just like two parallel SPDT switches combined into a single component with a single control. Both poles cannot be connected to a +ve throw. Such type of switches are used whenever a device is needed to be connected to a power supply with reversible polarity. DPDT switches often have a neutral position in which neither poles are not connected to their throws. Figure 6.4 shows the symbol of a DPDT switch. Figure 6.6 shows the connection of a DPDT with a motor.

Figure 6.5   Symbol of a DPDT switch

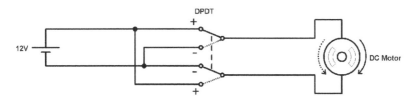

Figure 6.6   Connecting a DPDT switch with a motor

The most commonly used switches are manual switches (operated by hand). The other category, discussed later, is called Relay. Switches can also be classified according to their switching mechanism, the way in which they are turned ON (Closed) or OFF (Opened).

1. Push ON- Push OFF- Need to be pushed once to turn ON and pushed again to turn OFF (Bistable)
2. Push ON- Release OFF- Need to be pressed to turn ON and released to turn OFF (Stable OFF)
3. Push OFF- Release ON- Need to be pressed to turn OFF and released to turn ON (Stable ON)

i. Toggle Switches/Rocker Switches:

These switches come in the category of Push ON- Push OFF. Toggle switches have a lever which can be brought in to two positions, one connects the pole to the throw and the other disconnects them. Rocker switches have a button that can move like a seesaw to open or close the switch. They come as SPST, SPDT as well as DPDT switches. The symbol of a Toggle switch is shown in Figure 6.7.

Figure 6.7   Symbol of a Toggle/Rocker switch switch

ii. Push Buttons:

These switches come in all three switching mechanisms i-e Push ON- Push OFF, Push ON- Release OFF and Push OFF- Release ON. They come with two states only (SPST) and consist of a flat round top which is pressed to either connect or disconnect the pole with the throw. The circuit symbol of a pushbutton is shown in Figure 6.8:

Figure 6.8   Symbol of a push button

iii. Selector Switch:

This type of switch can be used to connect a single pole to multiple throws (two or more). It consists of a rotary knob which can elect one or more positions. This type of switch is used in a multimeter for selecting the range of the measured quantity. The symbol of a Selector switch is shown in Figure 6.9.

Figure 6.9   Symbol of a selector switch

Relays

An Electromechanical relay is a switch that uses an electromagnet to make or break contacts. Relays are mostly Single Pole Double Throw switches. As shown in Figure 6.10, a relay consists of a coil, a spring, an armature and 5 contacts, two of which are for the coil supply and 3 contacts form the SPDT switch. Like before, the pole is connected to one of the throws, whenever current is made to flow through the coil, it attracts the armature and connects the pole to throw 2, as soon as the current flow is stopped, the armature returns back to its original position because of the spring and connects the pole to throw 1 again. The coil can work with a DC as well as an AC supply.

Throw 2
Throw 1
Coil
Contacts
Pole

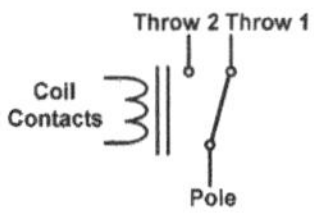

Figure 6.10 Symbol of a SPDT relay

One notable use of a relay is that it is used to switch a high voltage load with low voltages such as those coming from a computer. There are several types of relays such as Reed Relays, Solid State Relays and Latching Relay. Figure 6.11 shows some physical switches.

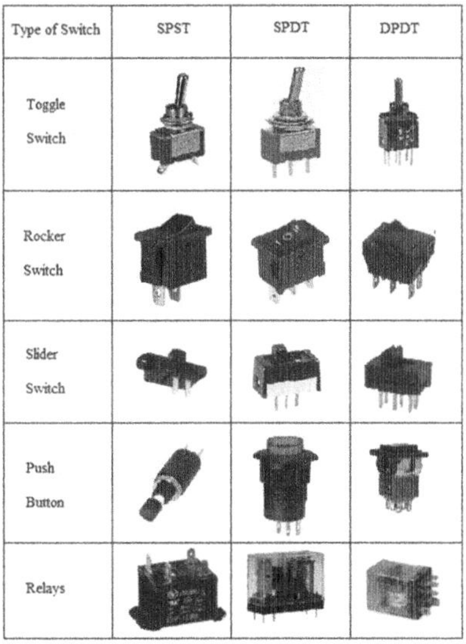

Type of Switch
SPST
SPDT
DPDT
Toggle Switch
Rocker Switch
Slider Switch
Push Button
Relays

Figure 6.11 Some real switches

Procedure

1. Take a Multimeter and set it on the buzzer function.
2. Take two pieces of wire and wrap them around the probe tips.
3. Perform the following tests and write down your answer in Table 6.1.

    i. Take an SPST, SPDT and a DPDT switch and name the terminals as A, B, C and so on.
    ii. Check connections among their contacts.

4. Take two SPDT Switches, a lamp and connectors for the mains power supply. Make connections as shown in the Figure 6.12.

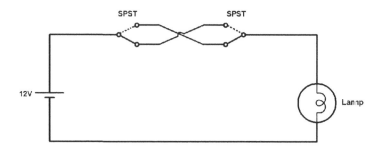

Figure 6.12   A two way switch

5. Take a 5V relay, a fan, an SPST switch and connectors for a power supply. Make connections as shown in Figure 6.13.

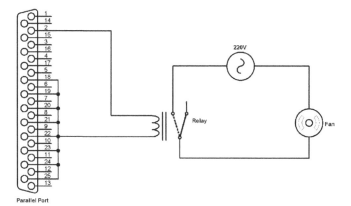

Figure 6.13   Motor control with a relay through a computer

6. Write down the following code in Turbo C and run it.

```
#include "conio.h"
#include "stdio.h"
#include "dos.h"

#define PORT 0x378
void main()
{
unsigned char c;
printf("Press O to turn the fan On and F to turn it off");
outportb(PORT,0x00);
while(!kbhit())
{
c=getche();
if(c=='O')
outportb(PORT,0x01);
else
if(c=='F')
outportb(PORT,0x00);
}
}
```

Observations

List the switches that you tested:

Table 6.1   Checking a switch

S N.o	Type of Switch	No of Stable states	Implication (Working or not)
1.			
2.			
3.			
4.			
5.			

Questions

1. Write down the working of the following types of switches:
i. Limit switch   ii. Proximity Switch   iii. Joystick Switch

_____

_____

_____

_____

2. What is switch bounce?

_____

_____

3. In what conditions does the lamp turn on for the two way switch?

_____

_____

Workshop # 07

Diodes

Object: To become familiar with Diodes

Apparatus:

1). A Digital Multimeter (DMM)
2). Few Diodes

Theory

All the components discussed so far (resistors, capacitors and inductors) were passive components. A diode on the other hand is an active electronic device that allows conduction in only one direction (within specified limits). It is constructed by fusing two different types of doped semiconductors (P-type and N-type) together. The word Diode is an abbreviation of Di-Electrode meaning a device with two electrodes i-e Positive (also called Anode) and Negative (also called Cathode). Figure 7.1 shows the constructional view and the symbol of a diode.

Figure 7.1   Constructional view and symbol of a diode

A junction is created at the point where the p-type and n-type material meet, this junction gives the diode its unique behavior. The voltage at which the diode starts to conduct is called the Barrier potential and it is dependent on the type of semiconductor and the impurity used to dope it, for silicon it is 0.7V and for germanium it is 0.3v.

There are two ways in which a diode can be connected to a power supply. First is Forward Bias (anode to positive terminal of battery and cathode to the negative terminal) in which current can flow from the anode to the cathode and the other is Reverse Bias (anode to negative terminal of battery and cathode to the positive terminal) in which the diode does not allow any current to flow through the by it as shown in Figure 7.2 and 7.3.

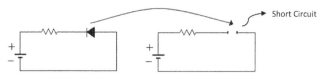

Figure 7.2   A diode connected in Reverse Bias

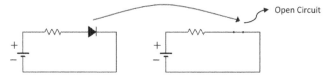

Figure 7.3   A diode connected in Forward Bias

For identification purpose, the cathode side of a diode has a colored ring near its end as shown in Figure 7.4. Another way to identify the terminals is by using the diode function on a multimeter, it will show a finite reading with the anode connected to the positive probe and cathode connected to the negative probe and infinity otherwise. Diodes are widely used for rectification purpose (conversion of AC into DC).

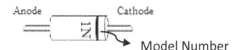

Figure 7.4   A real look of a diode

The Light Emitting Diode (LED) is a special type of diode that lights up whenever it is forward biased. Unlike the ordinary diode which is made using either Silicon or Germanium, an LED is made using Gallium Arsenide (GaAs) and some other semiconductor material compound that emits light. The semiconductor material determines the color of the light emitted by an LED and also the forward voltage required to light up the LED. LEDs are widely used for lighting purposes (ultra bright LEDs) and in displays. The symbol of an LED is shown in Figure 7.5.

Figure 7.5 Symbol of an LED

Like a diode, an LED needs to be forward biased for it to light up, in the reverse bias though, it behaves like an ordinary diode. The anode and cathode of an LED can be identified by using a multimeter set on the diode function or by making use of the fact that the anode of an LED is physically larger than the cathode. LEDs come in various sizes, 3mm, 5mm and 10mm. There are also bi-color LEDs having 3 terminals, which light up with one color with one polarity and with another for another polarity. These can either be made using an NPN or a PNP combination thus changing the polarities required to light them up. A special type of LED is the RGB (Red- Green- Blue) LED which has four terminals, one for each of the three colors and the remaining one for ground. There is also another type of LED that works in the infrared region of light (below the visibility range). This type is used in remote controls of televisions, air conditioners etc.

Procedure

1. Take a Multimeter and set it to the diode function.
2. Take a diode and check it using the method described in theory and fill in Table 7.1.
3. Repeat this for several diodes.
4. Take an LED and check it in the same way as the diode and fill Table 7.2.
5. Repeat this for several LEDs.

Observations

Table 7.1 Checking a Diode

S. N.o	Model	Voltage		Implication
		Reverse Bias	Forward Bias	
1.				
2.				
3.				
4.				
5.				

Table 7.2   Checking an LED

S. N.o	Color	On/Off		Implication
		Reverse Bias	Forward Bias	
1.				
2.				
3.				
4.				
5.				

Questions

1. What is meant by Barrier Potential?

   _____

   _____

2. What is meant by rectification?

   _____

   _____

3. Write down the names of some types of diodes?

   _____

   _____

Workshop # 08

Transistors

Object: To become familiar with the types of transistors and be able to check for their terminals

Apparatus:

1). A Digital Multimeter (DMM)
2). A few BJTs and FETs

Theory

A transistor is a 3 terminal device that is used in a variety of applications such as amplification and switching. There are two types of transistors categorized according to their construction:

1. Bipolar Junction Transistor
2. Field Effect Transistor

The Bipolar Junction Transistor (BJT) or simply known as a transistor comes in two flavors, NPN and PNP and depending on the type it is made up of a layer of extrinsic semiconductor sandwiched between two layers of the opposite type of semiconductor thus creating two junctions. A contact protrudes from each layer of the transistor and each has its own name. Figure 8.1 shows the schematic symbol and the construction of both types of BJTs.

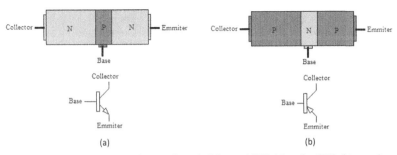

Figure 8.1 Constructional view and symbol for an NPN (a) and a PNP (b) transistor

45

The three terminals are named as Collector, Emitter and Base, as seen from Figure 8.1, the collector is the thickest of them all and it is also the most heavily doped, following it is the emitter which is moderately doped, the base is the thinnest and it is lightly doped. The base is common for the both the collector as well as the emitter. The difference between the NPN and the PNP transistors is the polarity of the supply required at their different terminals for operation. Since there are two junctions in the transistor (the base-emitter junction and the base-collector junction), we can make the diode equivalent of the transistors as shown in Figure 8.2.

(a)   (b)

Figure 8.2   Diode equivalent of a NPN (a) and a PNP (b) transistor

This allows us to check for the terminals of the transistor and also make sure if it's working or not by using the multimeter on the diode function. The voltage drop across the emitter-base junction would be slightly higher than the voltage the voltage drop across the base-collector junction. Terminal identification of transistors is necessary also because there is no standard sequence followed.

The second type of transistor is the Field Effect Transistor (FET). It is a two layer three terminal device made up of two pieces of semiconductors. Like the BJT, this comes in two flavors too, N-Channel and P-Channel. The construction and schematic symbol are shown in Figure 8.3.

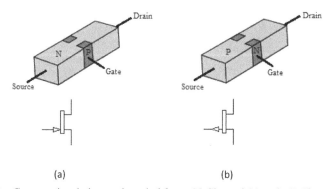

(a)   (b)

Figure 8.3   Constructional view and symbol for an N-Channel (a) and a P-Channel (b) JFET

The three terminals are called Gate, Source and Drain. The symbol and construction shown in Figure 8.3 are for the Junction Field Effect Transistor (JFET) in which there is a p-n junction between the gate and source, there is also another type of FET in which the gate is insulated from the source by means of a layer of SiO_2 called as the MOSFET (Metal Oxide Semiconductor Field Effect Transistor). Its construction and schematic symbol is shown in Figure 8.4.

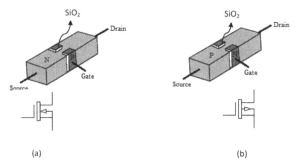

(a)   (b)

Figure 8.4 Constructional view and symbol for an N-Channel (a) and a P-Channel (b) MOSFET

Unlike other components discussed so far, transistors come in a housing also called a Package. Some types of packages available for transistors are shown in Figure 8.5.

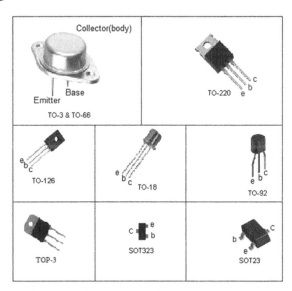

Figure 8.5 Different packages in which transistors are commercially available

Procedure

1. Take a Multimeter and set it on the diode function.
2. Take a Bipolar Junction Transistor and name the terminals A, B and C. Do the following and write down your observations in Table 8.1.

   i. Place the Red (+ve) probe of the DMM on A and the Black (-ve) probe on B. See if the DMM reading is infinite or finite.
   ii. Interchange the probe positions and observe the DMM reading.
   iii. Do this for all the combinations of probe positions for all three terminals, identify the three terminals of the Bipolar Junction Transistor and write the terminal names of A, B and C respectively. Also determine the type of the transistor (NPN and PNP). If the base is of N type, the transistor is NPN else it is PNP.

Observations

Table 8.1   Checking a BJT transistor

S N.o	Terminals of the BJT			Type of Transistor NPN/PNP	Implication
	A	B	C		
1.					
2.					
3.					
4.					
5.					

Questions

1. What is the difference between a PNP and an NPN transistor?
   _____
   _____

2. Write down the name of some of the packages in which transistors are available?
   _____
   _____

3. What is the difference between FETs and BJTs?
   _____
   _____

Workshop # 09

The Breadboard

Object: To become familiar with breadboards

Apparatus:

1). A Digital Multimeter (DMM)
2). Few hard wires (Gauge number 22)
3). A Breadboard

Theory

A breadboard also known as a Prototype Board or a Solderless Veroboard is a board for making and testing temporary electronic circuits so that if any component is connected incorrectly then it can be corrected. The board consists of several holes (also called sockets) arranged in rows and columns on a 0.1" grid, a typical breadboard is shown in Figure 9.1.

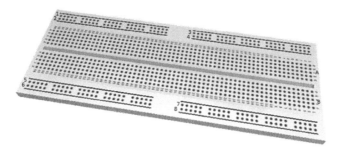

Figure 9.1 The breadboard

The number of rows and columns on a breadboard can vary. The breadboard shown in the above figure consists of two rows of holes (1, 2, 3, 4, 5, 6, 7 and 8) both at the top and the bottom, with two sets of 64 columns (A and B) in between. Each column has 5 holes which are connected to each other although there is no connection between adjacent columns. This is true for the rows too, wherein no two rows are connected and there is a disconnection between the two parts of them row. Components can simply be inserted in to the holes for connecting them together. Columns or rows can be connected to each other using a 0.6mm copper wire. Usually the two parts of each row are connected together so that they can be used as power rails.

Figure 9.2 gives a break out of the way holes are connected in a breadboard.

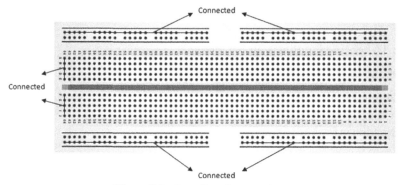

Figure 9.2   Breadboard connections

The connections between holes are accomplished through metal clips like the one shown in Figure 9.3.

Figure 9.3   Breadboard metal clip

These metal clips run through the breadboard underneath the hole mouths to connect each of the holes. Figure 9.4 gives a view of how columns are connected.

Figure 9.4 Inner view of a breadboard

Procedure

1. Take a Multimeter and set it on the buzzer function (Pg: 03).
2. Take two pieces of wire and strip them off at both ends.
3. Wrap the wires around the probes of the multimeter so that the other end sticks out.
4. Perform the following tests and write down your answer in Table 9.1.

    i. Pick a column and plug in one of the multimeter probes with wires on them in one of the holes. Now plug in the wire one by one in the remaining holes of the same column. See if they are connected or not.

    ii. Now having the multimeter probe plugged into the first hole, plug the other probe in the adjacent column. Check connectivity like before.

    iii. Repeat steps (i) and (ii) with other columns.

    iv. Now take out both the probes. Insert one of the probes in one of the holes in the rows, plug the other end in the adjacent holes, and check connectivity. Do this with all the holes in the same row half.

    v. While keeping the first probe in the row hole, plug the other probe in the other half and check connectivity.

    vi. Again keeping the first probe in the row hole, plug the other probe in another row and check connectivity.

5. Make the circuit shown in Figure 9.5 on a breadboard, and check its operation. One way of making connections on the breadboard is shown in Figure 9.6.

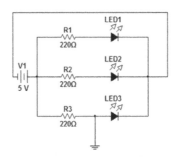

Figure 9.5   Circuit for connecting LEDs in Parallel

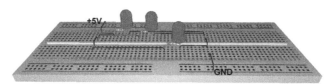

Figure 9.6   Breadboard connections for LEDs in parallel

6. Make the circuit shown in Figure 9.7 on a breadboard and check its operation. One way of making connections on the breadboard is shown in Figure 9.8.

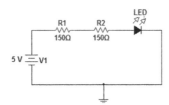

Figure 9.7   Circuit for connecting LEDs in series

Figure 9.8   Breadboard connections for LEDs in series

Observations

Table 9.1 Checking a breadboard for Shorts or Breaks

S N.o	Holes	Connected / Not Connected
1.	In the same column	
2.	In a different column	
3.	In the same half of row	
4.	In different halves of the same row	
5.	In different rows	

Questions

1. Write down one way in which can we connect two resistors on a breadboard (i) in parallel (ii) serially.

_____

_____

2. Was the breadboard you checked faulty or working properly?

_____

_____

Workshop # 10

Working with Resistors

Object: To learn to use resistors in circuits

Apparatus:

1). A Digital Multimeter (DMM)
2). Two 5KΩ Potentiometers
3). A 5KΩ Resistor
3). A Red LED
4). A Breadboard
5). Few hard wires (Gauge number 22)

Theory

We discussed about resistors in workshop 3 and learned how to decode their values from their color codes, now, we will use resistors in some very common but interesting applications. We all know about OHMs law i-e

$$V = IR$$

This law allows us to use resistors to alter the current and voltage in the circuit. First we will learn how to limit current using resistors.

Controlling the amount of current through a circuit is very important in a variety of situations such as lighting LEDs. For our discussion, we will consider the simplest case, calculating the current limiting resistor value for lighting up an LED. A Light Emitting Diode is a type of diode that lights up whenever it is forward biased. Different types of LEDs require different amounts of currents which also differ with its color. Let's take the example of a Red LED, a typical Red LED is rated for a voltage of 1.7 V and 15 mA. Consider the circuit shown in Figure 10.1.

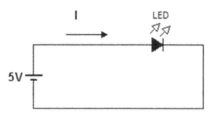

Figure 10.1   Lighting an LED with no current limiting

The LED in the above circuit would light up, but the problem with it is that the LED would draw a lot of current which could lead to it getting burned out.

A more efficient way to make up the above circuit is shown in Figure 10.2.

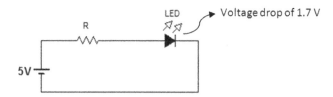

Figure 10.2   Limiting current for an LED

The Resistor R is used to control the amount of current. Taking the case of a Red LED requiring 15 mA of current, the value of the resistor can be calculated using Ohms law as follows:

$$R = \frac{V}{I}$$

Taking the LED has a voltage drop of 1.7v, the voltage V can be calculated as:

$$V = (voltage\ drop\ by\ LED)$$
$$V = 5 - 1.7$$
$$V = 3.33\ Volts$$

Now the value of the resistor can be calculated as:

$$R = \frac{33}{15\mathrm{x}10^{-3}}$$

$$R = 220\Omega$$

So one can use a 220Ω resistor in series with an LED to limit the current to 15 mA. One can use a similar procedure for finding out the resistance value for current limiting resistors for other applicatiosn. As mentioned above, resistors can also be used to limit the voltage. This is done with a very simple but useful circuit known as the voltage divider which is shown in Figure 10.3.

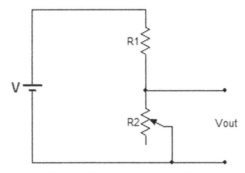

Figure 10.3   The voltage divider

The output voltage is given by the formula:

$$V out = R2 \text{ x } \frac{V supply}{R1 + R2}$$

As evident from the above equation, the output voltage is proportional to the resistance R2. The voltage divider is one of the most commonly used circuits in electronics, being used for scaling down voltages for Analog to Digital Converters, amplifiers, Televisions, Radios for setting volume, setting voltage references etc. A Light Dependent Resistor (LDR) is a special type of resistor whose resistance decreases with the increasing intensity of light falling on it.

Procedure

1. Take a Red LED, a 5V power supply, a 1kΩ potentiometer, two multimeters and a breadboard.
2. Set one of the multimeters to measure current and the other to measure resistance.
3. Make the circuit shown in Figure 10.4 on a breadboard. One way of making connections on the breadboard is shown in Figure 10.5.

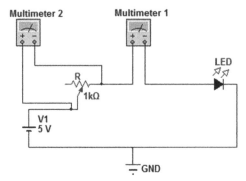

Figure 10.4   Circuit for using a resistor to limit current for an LED

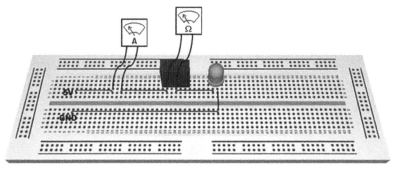

Figure 10.5   Breadboard connections for limiting the current for an LED using a resistor

4. According to the currents specified in Table 10.1, calculate the value of resistance required and set the potentiometer to the desired value. Note the current reading and fill in the table.

5. Take a resistor of 1kΩ, a potentiometer of 2kΩ, a multimeter and a 5V power supply and make connections as shown in Figure 10.6. One way of making connections on the breadboard is shown in Figure 10.7.

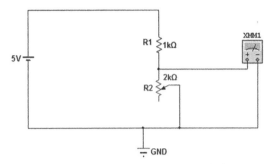

Figure 10.6 Circuit for the voltage divider

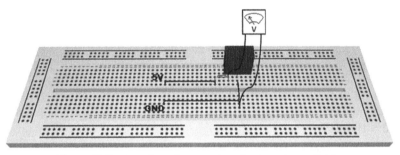

Figure 10.7 Breadboard connections for the voltage divider

6. Set the multimeter to measure voltage, vary the resistance of the potentiometer and measure the output voltage, fill in Table 10.2.

Observations

Table 10.1 Calculation of the current limiting resistance

Power Supply Voltage	Required Current	Calculated Resistance	Inserted Resistance	Measured Current Value
	10 mA			
	15 mA			
	25 mA			

Table 10.2   Calculation of the voltage divider output

Power Supply Voltage	R1	Potentiometer Resistance	Calculated Voltage Value	Measured Value

Questions

1. Will there be any difference in the current value if the current limiting resistance is wired after the LED?

2. Can the Voltage divider be used for AC voltage sources as well?

3. Using equal resistances for R1 and R2 in a voltage divider divides Vsupply by half. What is the difference in using R1 = R2 = 100Ω and R1 = R2 = 1000Ω?

4. Write down some applications of using voltage dividers in electronic circuits,

4. Take an LDR, a1KΩ potentiometer, a 5V Relay, an LED and a 220Ω Resistance to make a streetlight.

Workshop # 11

Working with Capacitors and Inductors

Object: To learn to use capacitors and inductors in circuits

Apparatus:

1). Function Generator and Oscilloscope
2). Few hard wires (Gauge number 22)
3). A Breadboard
4). Variable Capacitors
5). Two resistors of 1kΩ

Theory

As discussed before, capacitors and inductors combine with resistors and other components to perform some very interesting functions. This is because of the way the capacitors and inductors work. In this lab we will plot the phase shift that capacitors and inductors introduce in the circuit. In a capacitor, the current leads the voltage as shown in Figure 11.1.

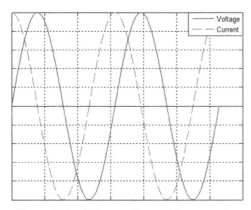

Figure 11.1   Voltage and current in a capacitor

Figure 11.1 shows that the phase difference is ideally going to be 90° and this phase difference causes a delay in the output voltage whenever an AC voltage is applied across a capacitor.

Similarly in the inductor too, there is a phase difference between the current and the voltage in which the current lags the voltage as shown in Figure 11.2.

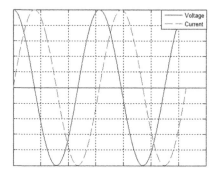

Figure 11.2   Voltage and current in an inductor

Like the capacitor, the phase difference is ideally 90° and this produces a delay in the current whenever an inductor is supplied with an (AC) voltage. Therefore one can use a capacitor and inductor to delay the output. A capacitor can be used along with a resistor for this purpose by using the connections shown in Figure 11.3.

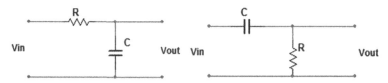

Figure 11.3   Delay circuits made by using a capacitor and resistor

Similarly one can make delay circuits using inductors as shown in Figure 11.4.

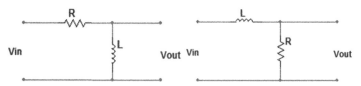

Figure 11.4   Delay circuits made by using an inductor and resistor

Procedure

1. Take a variable capacitor and a resistor of 1kΩ and make the circuit shown in Figure 11.5 on a breadboard. One way of making connections on the breadboard is shown in Figure 11.6.

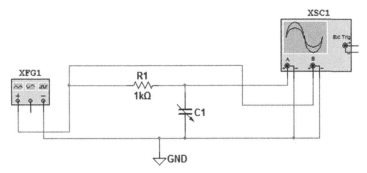

Figure 11.5   Delay circuit using a resistor and capacitor

Figure 11.6   Breadboard connections for the RC differentiator

2. Use the function generator to give the circuit a sine wave of 1Khz at point B, observe the shape of the wave on the oscilloscope, this is Vin, G is ground.
3. Using the other channel of the oscilloscope, observe the output at point A, this is Vout. Note the phase and amplitude difference between the input and output of the circuit.
4. Increase the capacitance, and observe the phase and amplitude difference between the input and output. Use Table 11.1 for setting capacitance values and fill in the appropriate column.
5. Take a variable capacitor and a resistor of 1kΩ and make the circuit shown in Figure 11.7 on a breadboard. One way of making connections on the breadboard is shown in Figure 11.7.

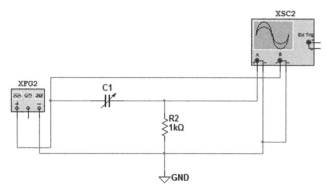

Figure 11.7   A capacitor resistor delay circuit

Figure 11.8   Breadboard connections for the RC integrator

6. Use the function generator to give the circuit a sine wave of 1Khz at point B, observe the shape and the phase of the wave on the oscilloscope, this is Vin.
7. Using the other channel of the oscilloscope observe the output at point A, this is Vout. Note the phase and amplitude difference between the input and output of the circuit.
8. Increase the capacitance, and observe the phase and amplitude difference between the input and output. Use Table 11.2 for setting capacitance values and fill appropriate column.

Observation

Table 11.1   Values of resistance and capacitance for the resistor capacitor delay circuit

S. N.o	Resistance	Capacitance	Phase Difference (°)
1.	1kΩ	0.083µf	
2.	1kΩ	0.166µf	
3.	1kΩ	0.332µf	
4.	1kΩ	0.885µf	
5.	1kΩ	0.133µf	

Table 11.2 Values of resistance and capacitance for the capacitor resistor delay circuit

S. N.o	Capacitance	Resistance	Phase Difference (°)
1.	0.083µf	1kΩ	
2.	0.166µf	1kΩ	
3.	0.332µf	1kΩ	
4.	0.885µf	1kΩ	
5.	0.133µf	1kΩ	

Questions

1. Why is the capacitor introducing a delay in the voltage?

_____

_____

2. What is the response of a capacitor and an inductor to a DC voltage?

_____

_____

3. What is the response of a capacitor and an inductor to AC voltage?

_____

_____

Workshop # 12

Working with Diodes

Object: To learn to use diodes in circuits

Apparatus:

1). A Digital Multimeter (DMM)
2). 2 5KΩ Potentiometers
3). A 5KΩ Resistor
3). A Red LED
4). A Breadboard
5). Few hard wires (Gauge number 22)

Theory

A diode conducts in only one direction while blocking flow in the other, this property can be used to convert AC in to DC. This process is known as rectification. Figure 12.1 shows a diode bridge which is made up of 4 diodes in 2 parallel arms and a pair of back to back diodes in each.

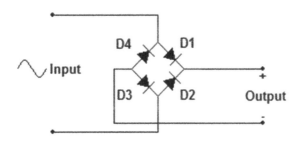

Figure 12.1   Diode bridge

Only one pair diodes conducts during each cycle and in this way AC is converted in to DC. Nowadays the bridge is available in a single package(IC), with four terminals protruding from it. One can use the diode function/Ohmmeter function on the multimeter to identify its input and output.

Procedure

1. Take a diode bridge and a Digital Multimeter.
2. Insert the bridge in a breadboard, and set the DMM to the option of Diode or Ohmmeter.
3. Name the terminals A, B, C, D.
4. Place the Positive (Red) probe on Terminal A and the Negative (Black) probe on terminal B. Note down the reading in Table 2.1. Interchange the DMM probes and fill the table.
5. Repeat the above step for all the terminals. Identify input and output according to the circuit schematic given in theory.

Observation

Table 12.1   Checking a diode bridge for faults

S N.o	Terminal				Resistance(Finite /Infinite)	Input Terminals	Output Terminals
	A	B	C	D			
1.	+	-	NC	NC			
2.	-	+	NC	NC			
3.	NC	+	-	NC			
4.	NC	-	+	NC			
5.	NC	NC	+	-			
6.	NC	NC	-	+			
7.	+	NC	NC	-			
8.	-	NC	NC	+			

Questions

1. Define rectification of an AC signal?

2. Why can the diode be used for rectification?

3. Enlist some applications of the diode.

4. Was the bridge you tested faulty or working properly?

Workshop # 13

The Veroboard and the Printed Circuit Board

Object: To become familiar with breadboards

Apparatus:

1). A Veroboard and a Copper Clad/Epoxy Board
2). Few hard wires (Gauge number 22)
3). A 12V Transformer
4). A Diode Bridge
5). A Capacitor of 5000μf
6). A Resistor of 1KΩ

Theory

A Veroboard also known as a Strip Board or Universal Board is a prototyping board that allows us to make permanent circuits. It consists of a grid of holes, each hole being 0.1" apart from the other. The bottom side of the veroboard is coated with copper which may connect the holes in a specific manner such as:

1. Each column of holes is separately (length wise) coated with copper thus connecting all the holes in each column
2. Each row of holes is separately (width wise) coated with copper thus connecting all the holes in each row
3. Each hole is coated with copper and thus no hole is connected to another hole(also called a Perf (Perforated) Board)
4. Various combinations of (1) and (2).

The copper path connecting the holes together is called a 'track', it is analogous to a wire for a breadboard. Wires are also used on a veroboard to connect holes not connected via tracks. A circuit is made by inserting the component leads from atop and soldering it from the bottom side. Soldering is the process of making an electrical connection between two metals by fusing them together using a third metal (Solder). It is done by using a soldering iron and some solder made up of tin and lead. Disconnections between holes, if required, can be done using a knife, razor or a drill bit. For soldering, the soldering iron is brought near the hole and the component terminal so that it touches them and a small amount of solder wire is pushed in to it, since the solder has a lower melting point than the copper as well as the terminal, it melts and fuses the hole and terminal together. The soldering iron is than retracted to allow the solder to cool down thus making the connection permanent. Mistaken connections can be corrected by touching the soldering iron to the fused joint and using a Desoldering pump to suck away the solder. The veroboard which comes in various standard sizes can be cut in to desired size using a junior hacksaw or a jewelers saw. A picture of a veroboard is shown in Figure 13.1.

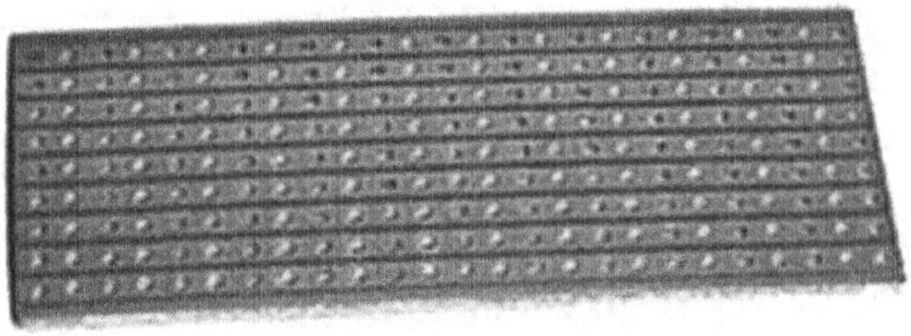

Figure 13.1   The veroboard

The Printed Circuit Board (PCB) is another method to make permanent circuits. It involves using a plain board with copper coated on either one side or both the sides. The circuit to be made is drawn on to the copper clad/epoxy board using permanent legible ink (using a pen or using the toner transfer method to transfer a design made using a computer, the latter one is more common nowadays). The copper is than etched (unwanted copper removed) away using a suitable etchant chemical (Ferric Chloride or a combination of Hydrogen per oxide and Acid) and warm water. After etching, only the part that was covered with the ink will be left untouched and all the residual copper will be removed. Holes can be made in the board using an Ejecto knife or a Mini Drill and components soldered in place. As for the veroboard, the lines of copper connecting the different holes together are called tracks. The PCB is used for mass production of electronic circuits. One disadvantage with the PCB is that once the circuit has been made, the connections cannot be changed.

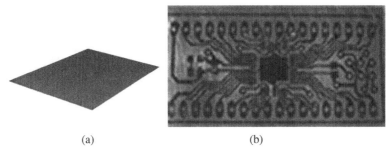

(a)   (b)

Figure 13.2 A blank PCB board (a) and an etched PCB board (b)

Procedure

1. Take a veroboard with its holes coated column wise. Make connections as shown in Figure 13.3.

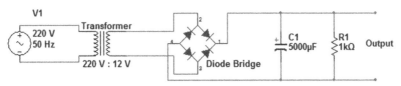

Figure 13.3 Circuit diagram of a diode bridge

2. Solder the components in place, give the supply to the labeled inputs and verify circuit operation.

Questions

1. What does the circuit discussed in the experiment do?

_____

_____

2. What is the difference between a Breadboard, Veroboard and a Printed Circuit Board?

_____

_____

3. What does one mean by tracks in a Printed Circuit Board?

_____

_____

4. Describe situations/stages of development in which the Breadboard, Veroboard and Printed Circuit Board are preferred to be used.

_____

_____

5. Briefly elucidate on the various methods used to create circuits on Printed Circuit Boards.

_____

_____

Workshop # 14

Introduction to Electronic Simulation

Object: Introduction to Simulation of electronic circuits.

Apparatus:

1). LT Spice
2). Computer with windows 98 or later.

Theory

Electronics simulation refers to the replication of the behavior of electronic devices using their mathematical models (equations governing their behavior). It allows us to design electronic circuits in software and test them for proper operation before moving on to make them in hardware, thus saving on cost and time.

There are numerous softwares available for the simulation of circuits. In this book we will use LTpice. LTSpice (Liner Technology Simulation Program for Integrated Circuits Emphasis), sometimes also refered to as SwiCad(Switcher Cad). LTSpice is a graphical version of SPICE(Simulation Program for Integrated Circuits Emphasis), incorporating a schematic/symbols editor and a SPICE simulator. The advantage of using spice simulation is that it allows us to model components as compared to other typical simulation softwares that have libraries with limited components.

When LTSpice is opened, we get the start up screen which is shown in Figure 14.1.

The screen consists of four main parts, the Main Menu (lets us create new files, open existing ones and manipulate the libraries, configure and perform circuit simulation), the Tool Bar (provides shortcut buttons for the common functions present in the main menu), the Status bar (tells us about the simulation i-e how much time has elapsed etc) and the Schematic Window (this is the area where the circuit is drawn). There is also a simulation window that pops up whenever a circuit is simulated. The A new file can be opened from File>New Schematic or by clicking the ⬚ icon as shown in Figure 14.2.

Figure 14.1 LTSpice start up screen

Figure 14.2 Creating a new schematic file

As seen in the Figure 14.2, several new options turn up in the main menu, these let us choose components, configure and perform simulations and carry out circuit analysis. One can use the various tools available to make and simulate circuits. Below is a summary of the available options. Each of these tools can also be selected from the Edit drop down menu.

Getting and working with the components

Components can be added using the ⟁ button, pressing *F2* or Edit>Component to open the add component dialog box, as shown in Figure 14.3.

Components can be searched by typing in their name. Once a component has been found, click on it and then click on *ok* to place the component in the schematic window by clicking at an appropriate place. Common components

Figure 14.3 Component selector dialog box

such as resistors, capacitors, inductors can be selected by using the, $\lessgtr$, $\doteq$ $\gtrless$ and $\doteqdot$ buttons on the tool bar, or by pressing *R*, *C*, *I* and *D* respectively. If a component has been placed incorrectly, it can be moved/dragged by using the , buttons on the tool bar, by pressing F7 and F8 for move and drag respectively. For deleting a component which is not required, one can use or hit the *Delete* button on the keyboard and click on the component. If one wants to change the name or value of any part, this can be done by right clicking over them. A dialog box pops up when this is done in which one can type in names and/or values.

Wiring the Components together

Before simulation, the components need to be connected together. The wiring tool can be used for this purpose. It can be selected by the button or by pressing *F3* on the keyboard. Wires are always straight lines. Turns can be made by clicking at the juncture and then moving in any direction, diagonal wires can be drawn by pressing the *control* key on the keyboard while dragging it diagonally. A node will automatically appear at a point where two or more wires connect. Wires can be labeled (named) using the button or pressing *F4* from the keyboard.

Simulating the Circuit

Once all the components are wired, the next step is to simulate the circuit and check the results. Before any kind of simulation is performed, the circuit needs to be connected to ground. This can be done using the ground tool, using the ⊽ button or by pressing *G*. Also one should make sure that there are no floating (unconnected) parts or wires in the circuit and that the components have correct values. For simulating the circuit, one can use the ⚡ button or click on Simulate>Run. When the circuit is first simulated, a dialog box will pop up requiring you to put in parameters for one of the six simulation types available as shown in Figure 14.4.

Figure 14.4 Choosing the type of simulation to be performed

The Transient analysis is used for simulation of any circuit for a certain period of time, AC Analysis is used for simulating the circuits response to an AC signal of increasing frequency, DC Sweep can be used for testing the circuit to an increasing DC voltage, Noise lets you calculate the amount of noise in the circuit due to Johnson, Shot and flicker noise, DC Transfer is used find the transfer function of a node voltage or a branch current, the last is the DC op pnt which can be used to calculate the DC operating point (bias point) of amplifiers etc. In this practical, we will limit ourselves to using the transient response. To stop the simulation in the middle, one can use the 🖐 button or click Simulation>Halt. The simulation can also be paused from the simulation drop down menu. The simulation results are shown in a separate window called the Waveform Viewer for AC sweep, DC sweep and Transient analysis. The results of the remaining two analyses are shown in dialog boxes. Voltages at nodes in the waveform viewer can be plotted by clicking on them (a voltage probe cursor appears 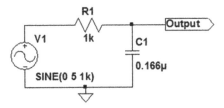 when this can be done), for measuring the current through any branch, one can click on any component in that branch (a current probe cursor appears when this can be done). Clicking on a component with the *Alt* key pressed plots the power dissipated by that component. The voltage difference between two nodes can be plotted by clicking on the first node and dragging the cursor to the other node without releasing the mouse button.

Other Spice directives such as parametric analysis etc can also be used to perform more custom analyses.

Procedure

1. Open a new schematic file and draw the circuit shown in Figure 14.5 in the schematic window.

Figure 14.5 Resistor capacitor circuit for simulation

2. Click on run and set the simulation parameters in the dialog box as shown in Figure 14.6.

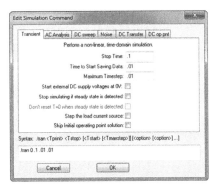

Figure 14.6 Simulation command parameters for simulation

4. Click on ok and plot the voltage at the output node and the input node and observe the phase difference.

Questions

1. Why does one do simulation of electronic circuits?

_____

_____

2. What is the purpose of the ⚡ button?

_____

_____

3. Describe the purpose of transient analysis?

_____

_____

4. Fill in Table 14.1 with the shortcut keys for the following functions.

Table 14.1   Shortcuts used in LTSpice

Function	Shortcut Key
Resistor	
Capacitor	
Inductor	
Diode	
Ground	
Wire	
Net Name	
Component Selector	
Rotate	
Mirror	
Delete	
Text	

Appendix A

Standard Resistor (5%) Values

1.0	10	100	1K	10K	100K	1M
1.1	11	110	1.1K	11K	110K	1.1M
1.2	12	120	1.2K	12K	120K	1.2M
1.3	13	130	1.3K	13K	130K	1.3M
1.5	15	150	1.5K	15K	150K	1.5M
1.6	16	160	1.6K	16K	160K	1.6M
1.8	18	180	1.8K	18K	180K	1.8M
2.0	20	200	2.0K	20K	200K	2.0M
2.2	22	220	2.2K	22K	220K	2.2M
2.4	24	240	2.4K	24K	240K	2.4M
2.7	27	270	2.7K	27K	270K	2.7M
3.0	30	300	3.0K	30K	300K	3.0M
3.3	33	330	3.3K	33K	330K	3.3M
3.6	36	360	3.6K	36K	360K	3.6M
3.9	39	390	3.9K	39K	390K	3.9M
4.3	43	430	4.3K	43K	430K	4.3M
4.7	47	470	4.7K	47K	470K	4.7M
5.1	51	510	5.1K	51K	510K	5.1M
5.6	56	560	5.6K	56K	560K	5.6M
6.2	62	620	6.2K	62K	620K	6.2M
6.8	68	680	6.8K	68K	680K	6.8M
7.5	75	750	7.5K	75K	750K	7.5M
8.2	82	820	8.2K	82K	820K	8.2M
9.1	91	910	9.1K	91K	910K	9.1M

Standard Capacitor Values

10pF	100pF	1000pF	.010μF	.10μF	1.0μF	10μF
12pF	120pF	1200pF	.012μF	.12μF	1.2μF	
15pF	150pF	1500pF	.015μF	.15μF	1.5μF	
18pF	180pF	1800pF	.018μF	.18μF	1.8μF	
22pF	220pF	2200pF	.022μF	.22μF	2.2μF	22μF
27pF	270pF	2700pF	.027μF	.27μF	2.7μF	
33pF	330pF	3300pF	.033μF	.33μF	3.3μF	33μF
39pF	390pF	3900pF	.039μF	.3μF	3.9μF	
47pF	470pF	4700pF	.047μF	.47μF	4.7μF	47uF
56pF	560pF	5600pF	.056μF	.56μF	5.6μF	
68pF	680pF	6800pF	.068μF	.68μF	6.8μF	
82pF	820pF	8200pF	.082μF	.82μF	8.2 μF	

Standard Inductor Values

1 nH/ μH	10 nH/ μH	100 nH/ μH	1000 nH/ μH
1.1 nH/ μH	11 nH/ μH	110 nH/ μH	1100 nH/ μH
1.2 nH/ μH	12 nH/ μH	120 nH/ μH	1200 nH/ μH
1.3 nH/ μH	13 nH/ μH	130 nH/ μH	1300 nH/ μH
1.5 nH/ μH	15 nH/ μH	150 nH/ μH	1500 nH/ μH
1.6 nH/ μH	16 nH/ μH	160 nH/ μH	1600 nH/ μH
1.8 nH/ μH	18 nH/ μH	180 nH/ μH	1800 nH/ μH
2 nH/ μH	20 nH/ μH	200 nH/ μH	2000 nH/ μH
2.2 nH/ μH	22 nH/ μH	220 nH/ μH	2200 nH/ μH
2.4 nH/ μH	24 nH/ μH	240 nH/ μH	2400 nH/ μH
2.7 nH/ μH	27 nH/ μH	270 nH/ μH	2700 nH/ μH
3 nH/ μH	30 nH/ μH	300 nH/ μH	3000 nH/ μH
3.3 nH/ μH	33 nH/ μH	330 nH/ μH	3300 nH/ μH
3.6 nH/ μH	36 nH/ μH	360 nH/ μH	3600 nH/ μH
3.9 nH/ μH	39 nH/ μH	390 nH/ μH	3900 nH/ μH
4.3 nH/ μH	43 nH/ μH	430 nH/ μH	4300 nH/ μH
4.7 nH/ μH	47 nH/ μH	470 nH/ μH	4700 nH/ μH
5.1 nH/ μH	51 nH/ μH	510 nH/ μH	5100 nH/ μH
5.6 nH/ μH	56 nH/ μH	560 nH/ μH	5600 nH/ μH
6.2 nH/ μH	62 nH/ μH	620 nH/ μH	6200 nH/ μH
6.8 nH/ μH	68 nH/ μH	680 nH/ μH	6800 nH/ μH
7.5 nH/ μH	75 nH/ μH	750 nH/ μH	7500 nH/ μH
8.2 nH/ μH	82 nH/ μH	820 nH/ μH	8200 nH/ μH
8.7 nH/ μH	87 nH/ μH	870 nH/ μH	8700 nH/ μH
9.1 nH/ μH	91 nH/ μH	910 nH/ μH	9100 nH/ μH

Appendix B

Some common prefixes and their factors

Prefixes	Factor
Femto(f)	10^{-15}
Pico(p)	10^{-12}
Nano(n)	10^{-9}
Micro(μ)	10^{-6}
Milli(m)	10^{-3}
Centi(c)	10^{-2}
Deci(d)	10^{-1}
Kilo(k)	10^{3}
Mega(M)	10^{6}
Gega(G)	10^{9}
Tera(T)	10^{12}
Peta(P)	10^{15}

Index

Authors Biography

Professor Dr BS Chowdhry is the Dean Faculty of Electrical, Electronics, Telecommunication and Computer Engineering at Mehran University of Engineering & Technology, Jamshoro (MUET), Pakistan. He did his BEng in Electronics from MUET in in 1983 and PhD from School of Electronics and Computer Science (ECS), University of Southampton, UK in 1990. He has more than 30 years of teaching, research and administrative experience in the field of Information and Communication Technology. He has the honour of becoming one of the editor of books "Wireless Networks, Information Processing and Systems", CCIS 20, and "Emerging Trends and Applications in Information Communication Technologies", CCIS 281, and "Wireless Sensor Networks for Developing Countries", CCIS 366, published by Springer Verlag, Germany. He is also Guest Editor of " Springer International Journal of Wireless Personal Communication". His list of research publication crosses to over 60 in national and international journals, IEEE and ACM proceedings. Also, he has Chaired Technical Sessions in USA, UK, China, UAE, Italy, Sweden, Finland, Switzerland, Pakistan, Ireland, Denmark, and Belgium. He is member of various professional bodies including: Chairman IEEE Communication Society (COMSOC), Karachi Chapter, Region10 Asia/Pacific, Fellow IEP, Fellow IEEEP, Senior Member, IEEE Inc. (USA), Senior Member ACM Inc. (USA).

Dr. Ahsan Ahmad Ursani is Professor and Chairman of the Department of Biomedical Engineering at Mehran University of Engineering and Technology Jamshoro, Sindh, Pakistan. He received his first degree in Electronic Engineering from the same university in 1995. He received his PhD from Institut National des Sciences Appliquées de Rennes, France, in 2008. He has been teaching courses in the undergraduate as well as postgraduate programs of Electronics, Telecommunication as well as Biomedical Engineering. His research interests include Medical Imaging, Image Processing, Speech Processing, Biomedical Signals processing, Biomedical Instrumentation, and Remote Sensing. From 2000 to 2006, he was a visiting scientist of the International Centre for Theoretical Physics, Italy. Apart from contributing several research papers in journals and conferences, he has contributed several popular science articles in his mother language, Sindhi. Dr. Ursani is actively engaged in making educational videos including those on Signals and Systems and other activities for the promotion of education, and especially for the promotion of education in mother language.

He has served as a reviewer for International Journal of Remote Sensing and Mehran University Research Journal of Science and Technology. He has also served as a member on the technical program committees of several international conferences.

Muhammad Zaigham Abbas Shah is a Lecturer in the Department of Electronics Engineering at Mehran University of Engineering and Technology, Pakistan. He has a undergraduate degree B Eng. in Electronic Engineering from Mehran UET and a postgraduate degree MSc. Electronic and Electrical Engineering from the University of Strathclyde, United Kingdom. His research interests include signal processing, image processing for monitoring applications and embedded system design.